SOLVING THE PYRAMID PUZZLE

Contents

Our Mysterious Planet

Sometimes it seems that there are few mysteries left in the world. Humans have walked on the Moon. Almost every inch of Earth has been mapped and explored.

But mysteries still remain. For example, there are still many questions about the ancient pyramids. Scientists have used many methods to look for the answers. No matter how hard they try, some questions still remain.

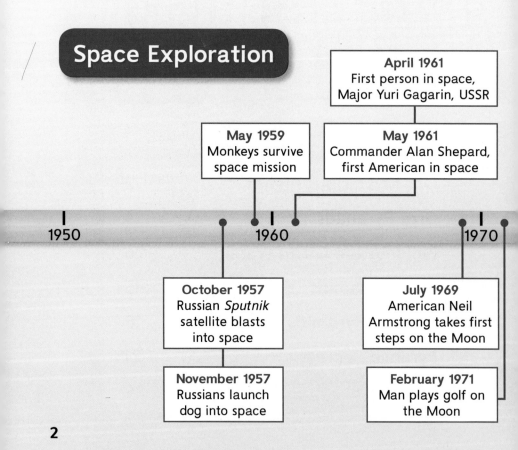

Space Exploration

May 1959
Monkeys survive space mission

April 1961
First person in space, Major Yuri Gagarin, USSR

May 1961
Commander Alan Shepard, first American in space

1950   1960   1970

October 1957
Russian *Sputnik* satellite blasts into space

November 1957
Russians launch dog into space

July 1969
American Neil Armstrong takes first steps on the Moon

February 1971
Man plays golf on the Moon

April 2001
First space tourist is billionaire businessman Dennis Tito

1980	1990	2000

January 2004
NASA rover looks for water on Mars

What does each interval of years on the timeline represent? How would the timeline change if the intervals were in 100s? 1,000s?

3

Giza Pyramids

Egypt's Giza (GEE zuh) pyramids are some of the most popular sites in history. As many as 750,000 tourists visit the pyramids each month.

Experts believe the pyramids were built as **tombs** for famous kings. The pyramids were built to last long after the kings had died.

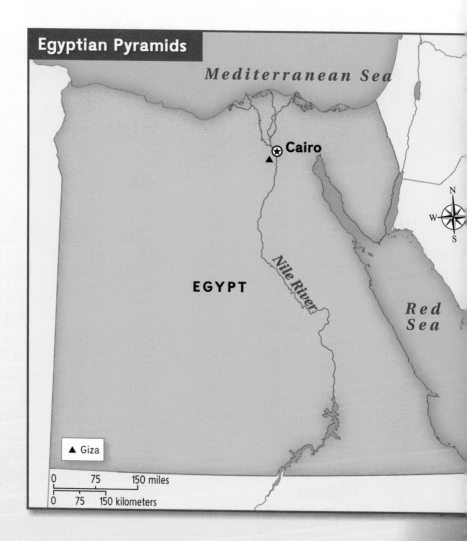

Egyptian Pyramids

Mediterranean Sea

⊛ Cairo

EGYPT

Nile River

Red Sea

N
W — E
S

▲ Giza

0   75   150 miles
0   75   150 kilometers

There is some disagreement about the age of the pyramids. Khufu's (KYOO fyoo) Pyramid, or the Great Pyramid, is the oldest and biggest. Scientists believe that the pyramid was built between 2589 and 2566 B.C.

Khafre's (KAH frah) and Menkaure's (mehn KYOO ray) pyramids were probably built during the next 100 years for Khufu's son and grandson.

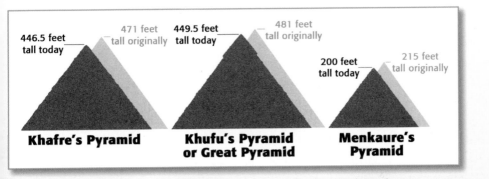

446.5 feet tall today

471 feet tall originally

449.5 feet tall today

481 feet tall originally

200 feet tall today

215 feet tall originally

Khafre's Pyramid

Khufu's Pyramid or Great Pyramid

Menkaure's Pyramid

Building the Giza Pyramids

Experts cannot believe how precisely the pyramids were built. There was no electricity. There were no heavy machines. The big pyramids were all built by hand.

The base of each pyramid is almost a perfect square. A square has four right angles. The angles of the Great Pyramid are almost exactly 90 degrees. Keep in mind that the base of the pyramid is 756 feet long on each side!

When it was finished, the Great Pyramid was 481 feet tall. This is almost as tall as a 50-story building. It was the tallest building in the world for more than 4,000 years.

Today, the Great Pyramid is $449\frac{1}{2}$ feet tall. Over time, about 31 feet of stone has crumbled from the point on top.

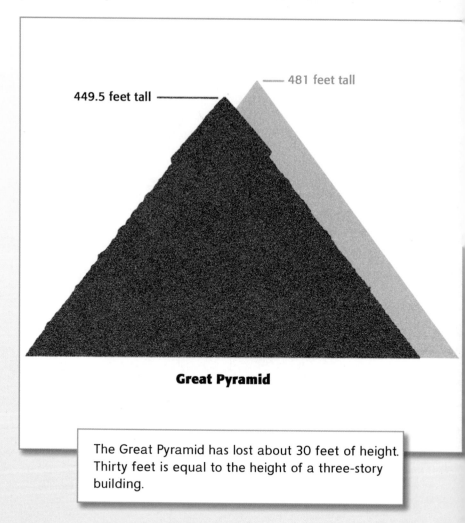

449.5 feet tall ————

———— 481 feet tall

Great Pyramid

The Great Pyramid has lost about 30 feet of height. Thirty feet is equal to the height of a three-story building.

The pyramids were built of huge stone blocks. A block of average size weighed $2\frac{1}{2}$ tons. A block was equal to the weight of $1\frac{1}{2}$ cars. Imagine moving the 2,300,000 blocks needed to build the Great Pyramid!

Did You Know?

The mortar (glue-like material) that was used to build the pyramids is made from a mysterious substance that scientists still do not know how to make.

The building blocks may have been moved with ramps, **pulleys,** and **levers**.

People have tried similar experiments. They wanted to see how long it would take to move a stack of blocks the same size as the ones used in the pyramids.

These experiments helped estimate how
many people worked on a pyramid. Scientists
believe that it took as many as 20,000 to 30,000
people to build these big pyramids.

These experiments also helped to find how long it might take to build the pyramids. It probably took about 20 years to finish each pyramid. There is no way to know if the estimates are right or wrong.

Experts work carefully to learn more about these old pyramids without hurting them.

People believed that the pyramid workers were slaves. Today, experts think that some workers were paid for their work. These workers built something big and lasting for their king and their country.

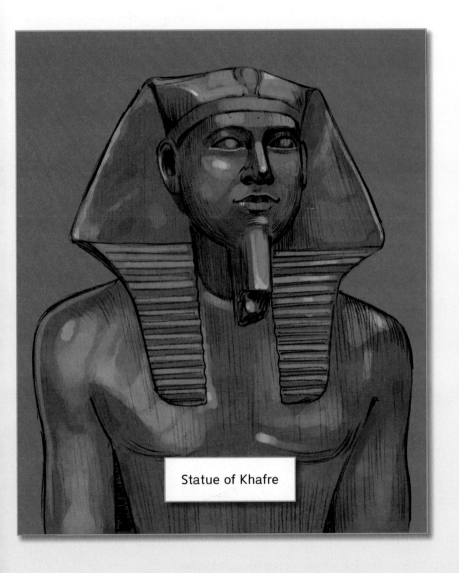

Statue of Khafre

Each face of the Great Pyramid was built at an angle of 51.5° degrees. The builders were careful that the angle did not change as the pyramid was built taller. If the angle changed, the four sides would not meet at the top.

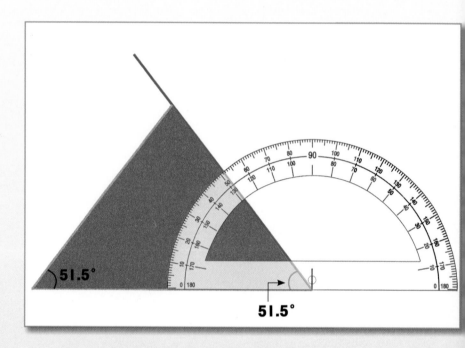

51.5°

51.5°

More Pyramids

Egypt is not the only place in the world where you can find pyramids. There are more pyramids in Central and South America than there are in Egypt.

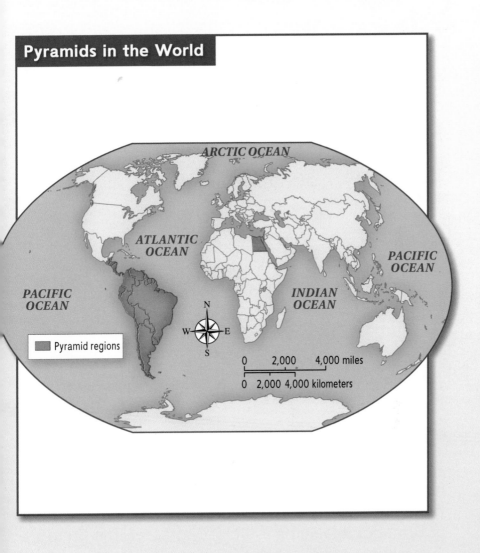

Pyramids in the World

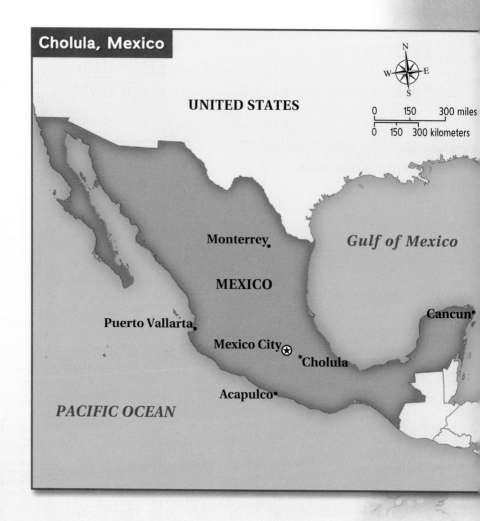

Cholula, Mexico

UNITED STATES

N
W E
S

0   150   300 miles
0   150   300 kilometers

Monterrey

Gulf of Mexico

MEXICO

Cancun

Puerto Vallarta

Mexico City

Cholula

Acapulco

PACIFIC OCEAN

The largest pyramid in the world is in Mexico. The base of this pyramid is a square with sides of 1,476 feet. Its height is 217 feet. It is not as tall as the Great Pyramid of Giza, but its volume is almost $\frac{1}{3}$ greater. The Great Pyramid of Cholula (choh LOOH lah) is also famous for being the largest **monument** ever built.

The Cholula pyramid was built slowly. Workers built the pyramid over 1,700 years.

The Cholula pyramid is also different because it has a flat top, not a pointed top. Long ago, a **temple** sat on the top of the pyramid.

Modern Pyramids

Today pyramids are still built. In Paris, France, a glass pyramid is an entrance to the Louvre (LOOHV) Museum. It stands 70 feet tall, and the sides of its square base are 115 feet in length. It is made of 603 rhombus-shaped glass panes and 70 triangular glass panes.

Louvre Museum in Paris, France

California is home to two pyramids. The Walter Pyramid is in Long Beach, CA. The shiny blue building is 18 stories tall. It was built as a sports arena for Long Beach State University.

TALK ABOUT IT

One story of a building is about 10 feet tall. How many feet tall is the Walter Pyramid? How many stories tall is the Louvre Museum?

The Transamerica Building is a famous San Francisco building. It looks like a tall, thin spike. It is 853 feet tall. The sides of the building's square base measure 152 feet in length.

Pyramid Secrets

The mysteries of the old pyramids may never be solved. Scientists keep studying these amazing structures and are learning more about the **cultures** that built them.

Technology keeps getting better too. Cameras and tiny robots let scientists see inside passages hidden deep in the pyramids.

Recent Discoveries

- In Egypt, a workers' village has been found that may have housed as many as 20,000 people.
- Old beds were found that suggest housing in the village.
- The workers ate a lot of beef, bread, and fish.
- The oldest coffin was found in a workers' cemetery. The coffin had been sealed for more than 4,500 years. Inside, a skeleton was found.

Each piece of information is a clue. Would you like to be a part of the team that solves the puzzle of the pyramids?

Picture of the Giza Pyramids taken from space.

Glossary

culture

The beliefs, customs, and social behavior of a particular group of people. *(page 21)*

lever

A bar that moves about a point and is used to move or lift a load. *(page 10)*

monument

A site or structure that is a tribute to a person, group of people, or an event. *(page 16)*

pulley

A tool made up of a rotating wheel and a rope or chain that is used to move heavy objects. *(page 10)*

temple

A building for worship. *(page 17)*

tomb

A burial place. *(page 4)*